EARTH SCIENCE

Landforms Around the World

MICHÈLE DUFRESNE

TABLE OF CONTENTS

What Is a Landform?	2
Continents and Oceans	5
Mountains and Hills	8
Plains and Plateaus	12
Valleys	15
Islands	18
Glossary/Index	20

PIONEER VALLEY EDUCATIONAL PRESS, INC

WHAT IS A LANDFORM?

Go outside and look around. What do you see? If you live in a city, you might see buildings, houses, or parks. Many of the places where we live were built by people. But beyond the city, you will find mountains, valleys, and ponds. You will find caves, volcanoes, and rivers. These were made by nature.

The earth is covered by different shapes. Some shapes rise high above the earth's **surface**, and others dip down into the ground. The different shapes of land and water that cover the earth are called landforms.

Some landforms were formed over many years. Wind, rain, and ice have worn away parts of the earth, changing the shape of the land.

Other landforms were made suddenly by natural disasters. An earthquake or an **erupting** volcano can quickly make a new landform.

CONTINENTS AND OCEANS

The two largest types of landforms are **continents** and oceans.

A continent is the largest **landmass** on the earth.

>> There are seven continents and five oceans on Earth.

If you looked at Earth from a spaceship, you would see that it is mostly blue. That is because there is more water than land on Earth.

Mountains, valleys, volcanoes, and other landforms can be found under the ocean too. Some underwater landforms are deeper, wider, and taller than any landforms found on land.

MOUNTAINS AND HILLS

Mountains are the highest landforms on Earth. They have steep sides and high peaks.

Many mountaintops are covered with snow all year long. Their peaks are too cold for most plants and animals to live.

The highest landform on Earth is Mount Everest. Every year, hundreds of people try to climb to its peak. The top of Mount Everest is called the "death zone" because so many climbers have died up there. The air at the top is thin and hard to breathe. The cold temperatures can also cause **frostbite**.

So why do people want to climb Mount Everest if it is so dangerous? Many people love the challenge of climbing this landform. One man who climbed it three times said he did it "because it is there." For climbers who love a challenge, standing on top of the world is a thrill they cannot resist.

Hills are much lower in height than mountains, but they are still higher than the land surrounding them. Hills are often covered with trees and plants.

There is a group of hills in Oregon called the Painted Hills. The colors of the hills change during different seasons and climates. They can be yellow, gold, black, or red. The different colors come from layers of different rocks. The Painted Hills were formed by volcanic eruptions.

PLAINS AND PLATEAUS

Plains are large, flat pieces of land with few trees. They make up more than one-third of the land on Earth.

Many of the world's farmers live on plains. The land has rich soil that is good for growing **crops** and raising animals. These areas are also called grasslands because of the tall grass that grows over them.

A very grassy plain is called a prairie. Prairie grass is kept healthy by being set on fire! The fire burns away weeds and other plants that can take water away from the grass. The ash from the burned plants makes the soil very rich. After a prairie fire, new grass grows back quickly.

A plateau is a flat, raised piece of land.

The sides of a plateau can be very steep.

Plateaus look like large mountains with flat tops.

They are found on every continent.

Some plateaus are formed by volcanoes.

VALLEYS

A valley is a low area of land between hills or mountains. The area at the bottom of a valley is called a floor, and the sides are called walls. Rivers or streams may flow across the floor of a valley.

A very narrow and deep valley is called a canyon. Canyons are formed when wind or water breaks off the sides of a mountain or plateau.

One of the most famous canyons in the world is the Grand Canyon.

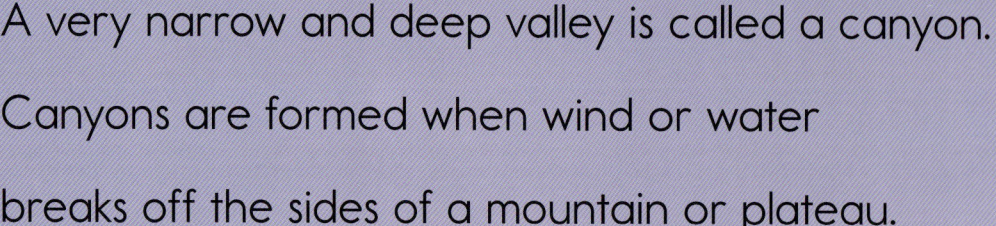

MORE TO EXPLORE

The Grand Canyon is so big that everyone in the world could fit inside it, and there would still be room for more people!

ISLANDS

An island is an area of land surrounded by water. Some islands are in the middle of lakes or rivers.

There are many islands in the ocean. Some of the best known islands are the Hawaiian Islands. They were formed by volcanoes that erupted in the ocean.

As the earth's temperature rises, ice around the world is melting. This pushes more and more water onto the shore. The world's islands are in danger of being flooded by rising water levels.

MORE TO EXPLORE

Some people living on small islands have to move their villages to new places because of rising seas and the wearing away of coasts because of **EROSION**.

Landforms

Marsh
wetland with grasses and few or no trees

Oasis
a place in the desert where there is water and some plants

Sinkhole
fallen soft rock that forms a hole on the earth's surface

Swamp
low, wet land with grass and trees growing in it

Waterfall
a place where running water drops over a cliff

Looking at

Cave
a large hole in the ground or a mountain with an opening to enter

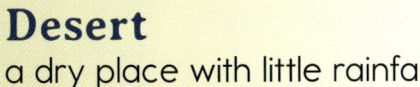

Desert
a dry place with little rainfall

Dune
a large pile of sand created by wind or waves

Geyser
a hole in the ground that spurts hot water and steam due to volcanic activity

GLOSSARY

continents
the largest areas of land on the earth

crops
plants that grow on a farm

erosion
the process by which something is worn away

erupting
sending out rocks, ash, and lava in a sudden explosion

frostbite
the freezing of parts of the body, such as fingers or toes

landmass
a very large area of land

surface
the upper layer of an area of land or water

INDEX

ash 13
canyon 16
caves 2
climates 10
continents 5, 14
"death zone" 9
disasters 4
Earth 3-6, 8-9, 12, 19
earthquake 4
erosion 19
erupting 4, 18
eruptions 10
fire 13
floor 15
frostbite 9
Grand Canyon 16
grasslands 12
Hawaiian Islands 18
hills 8, 10, 15
ice 4, 19
islands 18-19
landmass 5
layers 10
mountains 2, 7, 8, 10, 14, 15, 16
mountaintops 8
Mt. Everest 9
nature 2
oceans 5, 7, 18-19
Oregon 10
Painted Hills 10
peaks 8-9
plains 12-13
plateaus 12, 14, 16
ponds 2
prairie 13
rain 4
rivers 2, 15, 18
seasons 10
snow 8
soil 12-13
streams 15
valleys 2, 7, 15-16
volcanoes 2, 4, 7, 14, 18
walls 15
wind 4, 16